Protecting Agricultural Innovations

NIPA® GENX ELECTRONIC RESOURCES & SOLUTIONS P. LTD.
New Delhi-110 034

About the Author

Dr. Hitesh Kumar currently working as an Assistant Professor in the Department of Genetics and Plant Breeding at Banda University of Agriculture and Technology in Banda, Uttar Pradesh. Dr. Hitesh Kumar's childhood in village Jasoi, Distt, Muzaffarnagar, Uttar Pradesh, sparked a passion for agriculture. This passion led him to pursue Bachelor's degree in Agriculture followed by a Master's degree in Genetics and Plant Breeding from Ch. Charan Singh University in Meerut, Uttar Pradesh, and Ph.D. in Plant Breeding and Genetics from Punjab Agricultural University in Ludhiana, Punjab.

With 14 years of experience in teaching and research at the undergraduate and postgraduate level, Dr. Kumar is a leading expert in the agriculture field. He has also worked in the corporate sector at Syngenta India Ltd. in hybrid wheat breeding programme and has handled numerous externally funded research projects. As a co-guide, he has helped 18 students to earn their degrees and successfully guided four postgraduate students. Dr. Kumar's contributions to the scientific community are evident in his over 50 research papers published in peer-reviewed journals and eight book chapters contributed to books of national and international repute, showcasing his expertise in the field of Plant Breeding and Genetics.

Protecting Agricultural Innovations The Role of Intellectual Property Rights

Hitesh Kumar, Ph.D.
Assistant Professor
Department of Genetics and Plant Breeding
Banda University of Agriculture and Technology
Banda, Uttar Pradesh, India

NIPA® GENX ELECTRONIC RESOURCES & SOLUTIONS P. LTD.
New Delhi-110 034

NIPA® GENX ELECTRONIC RESOURCES & SOLUTIONS P. LTD.

101,103, Vikas Surya Plaza, CU Block
L.S.C.Market, Pitam Pura, New Delhi-110 034
Ph : +91 11 4386 0225, 9717133558, 9540816132
E-mail: newindiapublishingagency@gmail.com
Website: www. niparesources.com

Print ISBN: 978-93-95763-84-4
ebook ISBN: 978-93-58875-42-3

Composed and Designed by NIPA®.

Preface

Intellectual Property Rights (IPR) plays a significant role in promoting innovation and driving economic growth in countries, agriculture and allied sectors. Agricultural scientists and innovators rely on IPR protection to safeguard their rights and promote new and innovative practices in agriculture. However, despite the importance of IPR in the agriculture sector, there has been a lack of books covering the syllabus. To address this gap, we present book "Protecting Agricultural Innovations: The Role of Intellectual Property Rights". This book provides an easy-to-understand resource for undergraduate and postgraduate students interested in learning about IPR and its management in agriculture. With the help of the author's family and the support , this book, contains 11 chapters, aims to bridge the gap in the literature on IPR in agriculture.

I believe that this book will be a valuable resource for students and teachers in the agriculture discipline, as well as for graduate and postgraduate students in general Universities. It covers topics such as Patents, Trademarks, Copyright, Geographical indications (GI), Plant Breeder Right (PBR) and other related topics that are essential for anyone interested in understanding IPR and its role in agricultural innovation.

I welcome constructive feedback and suggestions from readers for further improvement in future editions, and we hope that this book will contribute to a better understanding and appreciation of IPR in agriculture.

Author

Contents

1

Introduction to Intellectual Property Rights: Concepts and Definitions

Introduction

Intellectual property (IP) refers to the creations of the human mind that can be protected and have commercial value. These creations, or "intellect," can take various forms, such as inventions, artistic works, and designs. Converting intellectual property into a useful product or process requires resources like time, effort, skill, and financial investment. Intellectual property rights (IPR) are legal rights granted to innovators and creators, giving them exclusive control over the use and monetization of their inventions and creations. IPRs are territorial and allow creators to sell, license, transfer, or otherwise dispose of their rights, similar to physical property.

IPRs are considered crucial for promoting economic growth and development, as they encourage innovation, investment, and entrepreneurship. Without IPRs, inventions and creative works could be easily copied or stolen, resulting in reduced incentives for innovation and decreased economic value. Lack of IPR can also lead to an increase in infringements and economic losses. In recent years, India has been increasing its awareness of IPR and has seen a rise in registrations of indigenous technology, patents, trade secrets, and copyrights among others.

With the evolution of agricultural products and raw materials as global commodities traded on a large scale, promoting the dissemination, commercialization, and further improvement of agricultural inventions and technologies has become crucial in today's interconnected world. Therefore, it is essential to recognize the importance of intellectual property rights (IPR) in agriculture and allied sectors, which will facilitate the sharing of knowledge encourage innovation, and enable creators to derive economic benefits from their inventions and creations.

Property meaning

Property refers to items that are legally owned by individuals or companies which can be acquired through purchase or gifting. Examples of property include physical items like automobiles, industrial machinery, furniture, electronics, land etc. There are two main types of property:

a. **Tangible property:** This type of property is physical and can be measured or touched. Examples include land, vehicles, equipment, machinery, furniture, inventory, securities like stocks and bonds, and cash. Tangible property can be further classified into two subcategories:

i. **Current assets:** These are assets like bonds and fixed deposits that can be easily converted to cash.

ii. **Fixed assets:** These are assets like trucks, machinery, equipment, office furniture, buildings, land, and farm implement that are used for long-term operations.

b. **Intangible property:** This type of property is not physical but has value in the future. Examples include patents, copyrights, trademarks, and trade secrets.

Property can also be classified based on its mobility as follows:

a. **Immovable property:** This type of property cannot be moved and is fixed in one location. Examples include houses, factories, and manufacturing plants, which are protected by land laws of the country.

b. **Movable property:** This refers to property that can be moved from one place to another and is not fixed in one location. Examples include computers, animals, jewelry, vehicles, farm machinery, gold, silver, money, and furniture.

Historical Development of Intellectual Property Rights: The historical development of Intellectual Property Rights (IPR) can be traced back to ancient times, as evidenced by special marks on pots during the Harrappa civilization. Initially, countries such as Germany and America had their own laws for protecting intellectual property. However, it was later realized that there was a need for international understanding and harmonized laws among countries to protect IP. The Paris Convention marked the beginning of this understanding among member countries. In 1893, the United International Bureau for Protection of Intellectual Property (BIRPI) was established in Berne, Switzerland.

After World War II, the economies of many European and Asian countries were devastated. To revive their economies, the United Nations established three bodies in 1947: the World Trade Organization (WTO), the International Monetary Fund (IMF), and the International Trade Organization (ITO). The General Agreement on Trade and Tariff (GATT) was created with the purpose of increasing predictable, stable international trade, reducing tariffs, settling disputes among countries, and holding regular negotiations among member countries. India became a member of GATT in 1947, allowing for the import of useful domestic goods such as oils, industrial raw materials, and new technologies, as well as the export of indigenous products.

With the increasing awareness and need for IPR, the BIRPI was shifted from Bern to Geneva in 1960 and renamed the World Intellectual Property Organization (WIPO) in 1967. WIPO's objective is to promote the creation of human mind, invention for economic, social, and cultural progress through international cooperation. It is responsible for setting norms, standards, executing legal and technical assistance, and registering IP among member countries.

The recognition of agriculture as a rule-bound enterprise of investment and profit-making in the General Agreement on Tariffs and Trade (GATT) for the first time was in the Uruguay Round (1986-1994). The round led to the establishment of the World Trade Organization (WTO) in January 1995. WTO has at least half a dozen intergovernmental agreements that directly affect agriculture, including the Agreements on Agriculture (AoA), Applications of Sanitary and Phytosanitary Measures (SPS), Technical Barriers to Trade (TBT), Anti-Dumping, Subsidies and Countervailing Measures and Trade-Related Aspects of Intellectual Property Rights (TRIPS).

The TRIPS agreement provides for internationally acceptable and enforceable patent protection for new inventions in all areas of technology. It is the most comprehensive international agreement on intellectual property. The WTO has also issued a statement clarifying the scope of TRIPS "to promote access to technology for all." In India, inventions were protected from 1856 based on Act VI of the British Patent Law of 1852. As per need-based modifications were adopted several times, and the act was re-amended in 1911 as the Indian Patent and Design Act. Further amendments were made, and the act was renamed the Indian Patent Act 1970.

Intellectual Property

Intellectual property (IP) refers to the creations of the mind used in commerce, such as inventions, literary and artistic works, designs, symbols, names, and images. Legal protection is available through various means, including patents, copyrights, trademarks, and trade secrets, which allow individuals and companies to benefit financially from their innovations while maintaining exclusivity. The IP system aims to balance the interests of innovators and the public by fostering an environment in which creativity and innovation can thrive. Intellectual property rights (IPR) grant exclusive legal rights to innovators and creators for the sale, marketing, transfer, licensing, and other forms of economic benefits related to their inventions and creations. The government provides these rights for a specific period, subject to defined terms and conditions, Examples of IPR including copyrights, patents, trademarks, industrial designs, indications, trade secrets, and layout designs on integrated circuits.

Innovation, invention, and creativity are essential for the growth of organizations, and intellectual property (IP) plays a crucial role in protecting and incentivizing such activities. IP can be anything from a new renewable energy technology to a novel, a painting, a logo, or a brand name, and legal protection of these rights encourages investment in research and development, leading to technological advancements and the creation of new products and services. The benefits of IP are significant for both individuals and society, as it promotes economic growth and job creation while ensuring that new products and services meet quality standards and protecting consumers from substandard or dangerous products. In summary, IP is a critical aspect of modern commerce promoting creativity, innovation, and economic growth while balancing the interests of innovators and society as a whole.

Types of Intellectual Property Rights

Intellectual property can be classified into different types based on their nature, application, and characteristics. Here are some examples of each type:

a. **Copyright:** Copyrights protect the expression of ideas by authors, artists, and creators. It safeguards the form of expression rather than the subject matter of the writing. This type of intellectual property covers literary and artistic works, musical works, photographic works, motion pictures, computer software, performing art, and broadcasting work.

b. **Patents:** Patents are rights given by the government to inventors or creators for inventions that meet certain criteria, such as global novelty, non-obviousness, and industrial application. It provides the exclusive right to the inventor to prevent others from making, using, or selling the invention for a certain period of time.

c. **Trademarks:** Trademarks refer to any sign or logo on a product that identifies it as produced by a specific person, company, or institute. It is a symbol of quality, recognition, reputation, and trust among the customers.

d. **Industrial designs:** Industrial designs cover the creativity of making the model or aesthetic appearance of a product. It refers to the shape, configuration, pattern, or ornament of the product. The design will decide the product market appeal for selling. For example, watch, car, home appliance design, etc.

e. **Geographical indications:** Geographical indications identify goods that have certain quality parameters influenced by the local climate, soil, or traditional knowledge. For example, agricultural goods, foodstuff, textile, handicrafts, etc.

f. **Trade secrets:** Trade secrets refer to any invention or knowledge that has business use for economic benefit and is kept confidential. This type of intellectual property is not patentable. Examples include recipes, designs, formulas, etc.

g. **Layout design of semiconductor integrated circuit:** The home appliance and electronic gadget are controlled by a small integrated circuit and microprocessor. These circuits are developed by skilled engineers and protected under the Layout Design of Semiconductor Integrated Circuit Act.

h. **Plant variety protection:** This type of intellectual property is used to protect new varieties of plants that have been developed through selective breeding or genetic engineering. It gives the owner exclusive rights to market and sell the plant variety for a certain period of time.

i. **Data exclusivity:** This form of intellectual property refers to the protection of clinical trial data that is submitted to regulatory agencies for approval of drugs and medical devices. It prevents generic drug manufacturers from relying on the data to seek approval for their products for a certain period of time.

j. **Domain names:** A domain name is a unique identifier that is used to locate a website on the internet. While not strictly a form of intellectual property, domain names can be protected through trademark law if they are used to identify and promote a brand.

k. **Utility models:** This type of intellectual property protection is similar to patents but is often used for minor inventions or improvements to existing products. It provides a shorter period of protection than a patent, typically around 10 years, and has less stringent requirements for novelty and inventiveness.

Intellectual Property

The Intellectual Property (IP) covers intangible creations of human intellect, such as ideas, inventions, literary and artistic works, symbols, names, and images used in commerce, which possess value. It is a legal framework that provides protection for creative and innovative works, including inventions, innovations, creativity, and discoveries. Therefore, IP can refer to a variety of works such as inventions, innovations, creativity, and discoveries.

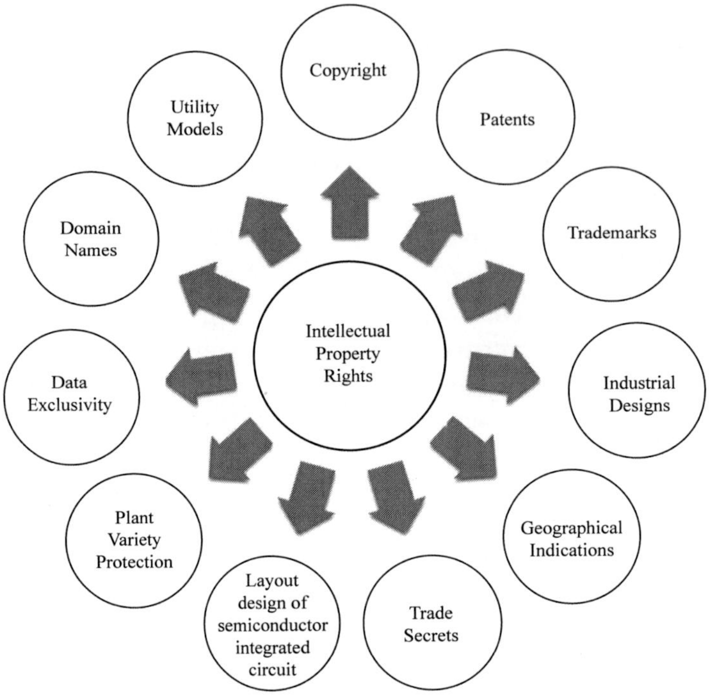

Different forms of Intellectual Property Rights

a. **Invention:**Invention is the process of creating a new product or process that has never existed before. It typically involves developing a unique function that may lead to a significant breakthrough. Examples of inventions include scientific and technological advancements such as railroads, aviation, vaccinations, antibiotics, astronautics, computing internet, and smartphones. Socio-political inventions such as the Red Cross parliaments, and constitutions also fall under this category. Additionally cultural inventions like painting, printmaking, photography, cinema and various musical genres (e.g. rock, opera) can also be considered inventions. Inventions are often novel and non-obvious, and they can have a significant impact on society and the economy.

b. **Innovation:**Innovation is the process of taking a new idea or product and turning it into a commercially viable and widely used solution. It involves the implementation of new features or modifications to existing products to meet the changing needs of consumers or businesses. Innovation include the introduction of smartphones, which are an improvement on traditional mobile phones, and the development of electric cars as an alternative to gasoline-powered vehicles and implementation of online shopping platforms, which revolutionized the way people buy goods and services.

Innovation is the proper utilization of creativity and invention, as it involves taking an existing idea and adding value to it to create something new and useful. Through innovation, businesses and individuals can improve products and services, enhance efficiency, and gain a competitive edge in the marketplace.

c. **Creativity:** Creativity is the ability to generate unique and original ideas and the skill to turn them into reality. It involves using imagination, critical thinking, and problem-solving skills to come up with new and innovative solutions. For example, a designer who creates a new fashion line, an artist who paints a unique piece of artwork, or an engineer who develops a new machine are all examples of creativity in action. Creativity is not limited to the arts or sciences, but can be applied to any field, including business, technology, and social innovation. It is an important trait for entrepreneurs and innovators, who are constantly seeking new and better ways to solve problems and meet customer needs.

d. **Discovery:**Discovery involves the identification or unearthing of something new or previously unknown. This could be a new species, a new planet, a new chemical element, or a new law of nature. Discoveries are often made through scientific research, exploration, or experimentation. They can also be made accidentally, such as the discovery of penicillin by Alexander Fleming. In contrast to inventions, discoveries are not created, but rather are found or revealed. For example, the discovery of the structure of DNA by Watson and Crick was a groundbreaking discovery that has had a significant impact on the field of genetics. Similarly, the discovery of the Higgs boson particle by scientists at CERN was a major breakthrough in particle physics.

Attribute of IPR: The following are the main features of Intellectual Property Rights (IPR).

a. **Measurable Unit:** IPR can be measured in the form of an idea, process, product, invention, or innovation developed by a person.

b. **Input:** Developing IPR requires inputs such as time, skill, planning, thinking, and financial assets.

c. **Reproducibility:** IPR can be copied, multiplied, and sold, which can result in a loss for the inventor or creator. Therefore, legal protection is required to encourage inventors.

d. **Uniformity:** The product form of IPR should be similar when it is commercially exploited.

e. **Exclusivity:** The product form of IPR should have a unique feature, trait, or quality parameter that distinguishes it from other products available in the market.

f. **Transferability:** Transferability refers to the transfer of rights of the inventor for their invention to a third party for utilization in business purposes.

Importance of IPR

The importance of Intellectual Property Rights (IPR) lies in their ability to encourage and protect innovation and creativity. Here are some reasons why IPR is important.

a. **Innovative ideas as means to earn profit** – Ideas themselves may have little to no value. However, by utilizing intellectual property rights, innovators can turn their ideas into commercially viable products or services. Registering for copyrights or patents can result in a constant stream of fees and increased income, ultimately boosting the overall market results.

b. **Export business opportunities** – Intellectual property can also improve a company's productivity in the export market. By using logos or designs protected by IP rights, companies can sell their products and services in foreign countries and potentially obtain franchise arrangements with overseas corporations, or export their proprietary products.

c. **Encouraging innovation through securing intellectual property** – Despite having a unique idea or development, there will always be people who attempt to duplicate it for monetary gain. Therefore, protecting IP rights is crucial to prevent any third-party infringements. IP protection can be implemented for businesses of all sizes, and it's essential to evaluate market needs and situations to determine which Intellectual Property Rights (trademark, copyright, or patent registration) to use in covering various areas of IP.

d. **Business growth** – Shielding exclusive goods or services from competitors is crucial for small-scale enterprises to sustain growth and profits. Losing market share at an initial level can be harmful to a company's long-term health, making IP protection a critical aspect of business strategy.

Why to Protect IPR

a. Protection of IPR is important to prevent the unauthorized use of intellectual property by others without the permission of the rightful owners.

b. Protection of IPR encourages innovations and creativity by providing incentives to inventors and creators through legal protection.

c. Protection of IPR promotes investments in research and development as companies and individuals are more likely to invest in new technologies and ideas if they can protect their intellectual property.

d. Intellectual property can be traded as goods and services, and the protection of IPR ensures that the rightful owners can receive compensation for their creations.

e. Protection of IPR contributes to national economic prosperity by encouraging innovation, promoting entrepreneurship, and creating new jobs and industries.

Nature of IPR

a. **Monopoly rights:** Intellectual property rights grant monopoly rights to the owner for a certain period of time, enabling them to exclude others from using or exploiting their invention, design, or creative work without their permission. This exclusive right allows the owner to control the use, production, and distribution of their intellectual property.

b. **Territorial rights:** Intellectual property rights are generally granted on a territorial basis, which means that they are valid only in the country or region where they are granted. If a person or business wants to protect their intellectual property in other countries, they must apply for separate protection in each country or rely on international agreements, such as the Paris Convention, which provides some level of protection in member countries.

c. **Fixed term:** Intellectual property rights have a fixed term of protection, which varies depending on the type of intellectual property. Patents, for example, have a fixed term of 20 years from the date of filing, while trademarks and copyrights can last indefinitely as long as they are renewed and remain in use.

d. **Renewed from time to time:** In order to maintain protection, intellectual property rights must be renewed from time to time, either by paying renewal fees or by meeting other requirements set by the governing body. Failure to renew the intellectual property right can result in the loss of protection.

e. **Simultaneously held in many countries:** Intellectual property rights can be held simultaneously in many countries, provided that the owner applies for protection in each country where they want to have their intellectual property protected. International treaties and agreements, such as the

Madrid Protocol, make it easier for owners to seek protection in multiple countries at once.

f. **Assigned, gifted, sold, and licensed:** Intellectual property rights are treated like any other form of property and can be assigned, gifted, sold, or licensed to other parties. For example, an inventor can assign their patent rights to a company in exchange for royalties or a lump sum payment, or a songwriter can license their copyrighted music to a film production company for use in a movie.

Applications of IPR in Agriculture

The agriculture and allied sectors are vital sources of food, rural employment, and raw materials for the industrial sector. Thanks to globalization and technological advancements, these sectors have made significant progress towards economic growth, poverty alleviation, and the supply of food and raw materials. In this context, intellectual property rights (IPR) have become increasingly important as they offer potential economic benefits. There are various ways in which IPR is applied in agriculture and allied sectors.

i. Plant Breeder's Rights (PBR) give exclusive rights to plant breeders to control the production, sale, and distribution of their new plant varieties.
ii. PBR protection stimulates plant breeders to develop new crop varieties that are more resistant to diseases, pests, and environmental stresses, resulting in higher yields and improved quality.
iii. Patents protect new inventions, including biotechnological inventions like genetically modified crops.
iv. Patents provide a significant return on investment for companies that invest heavily in research and development while also encouraging technology transfer and collaboration.
v. Geographical Indications (GI) protection is used to safeguard the names of products that are linked to a specific geographic region and have unique characteristics or qualities.
vi. GI protection helps prevent the unauthorized use of unique names and can provide economic benefits to farmers and producers in the region.
vii. Trade secrets are particularly important in the case of proprietary technologies that offer a competitive advantage.
viii. Trade secrets protect confidential information, such as formulas or processes used in the production of agricultural products.
ix. Copyright protection is critical for publications and software used in agriculture, like agricultural textbooks and farm management software.

x. The use of IPR protection in agriculture and allied sectors encourages innovation, technology transfer, and collaboration.

IPR protection is critical in promoting sustainable agricultural practices and can provide economic benefits to innovators and investors in agriculture and allied sectors. PBR, patents, GIs, trade secrets, and copyrights are some of the IPR tools that can be used to achieve these goals. The protection of intellectual property rights encourages and supports innovation, collaboration, and technology transfer, ultimately leading to the development of better and more efficient agricultural practices.

2

Intellectual Property Governing Bodies Agreement and Treaty

Intellectual property (IP) refers to creations of the human mind, such as inventions, literary and artistic works, symbols, names, images, and designs, which are protected by law. The protection of intellectual property is essential for promoting innovation, creativity, and economic growth. In order to ensure effective protection of intellectual property rights, various governing bodies, organizations, agreements, and treaties have been established at national and international levels. These bodies are responsible for setting and enforcing standards and regulations related to IP, as well as facilitating cooperation and coordination among different countries and stakeholders. Some of the major international organizations and agreements that deal with intellectual property include the World Intellectual Property Organization (WIPO), The Agreement on Trade-Related Aspects of Intellectual Property Rights (TRIPS), and the Paris Convention for the Protection of Industrial Property. These organizations and agreements play a crucial role in creating a conducive environment for innovation, investment, and development, while at the same time ensuring fair and equitable access to intellectual property for all.

The major International Intellectual Property Organizations are

i. **World Intellectual Property Organization (WIPO):** The World Intellectual Property Organization (WIPO) is a specialized agency of the United Nations (UN) that was estabclished in 1967 with the goal of promoting and protecting intellectual property (IP) rights worldwide. It has its headquarters in Geneva, Switzerland. WIPO has 193 member countries making it one of the largest and most diverse international organizations in the world. Membership in WIPO is open to all UN member states, as well as some intergovernmental organizations and regional economic integration organizations. WIPO's member countries work together to promote and protect IP rights, share best practices, and address global IP challenges. WIPO provides a range of services and resources to support the development, use, and protection of intellectual property rights. Some of its key functions include :

a) Developing international IP laws and norms
b) Providing legal and technical assistance to member countries
c) Administering international IP treaties and agreements
d) Offering training and capacity-building programs on IP issues
e) Conducting research and analysis on IP-related topics
f) Maintaining databases and other resources on IP-related information

ii. **World Trade Organization (WTO):** The World Trade Organization (WTO) is an intergovernmental organization that was established on January 1, 1995, following the signing of the Marrakesh Agreement. The WTO headquartered is located in Geneva, Switzerland and is responsible for administering and enforcing international trade agreements between member countries. The WTO has 164 member countries , including major economies such as the United States, China, Japan, and the European Union. Some of its key functions include :

a) The WTO administers various trade agreements negotiated between member countries, including the General Agreement on Tariffs and Trade (GATT) and the Agreement on Trade-Related Aspects of Intellectual Property Rights (TRIPS).

b) The WTO provides a forum for member countries to negotiate new trade agreements and to resolve disputes related to existing agreements.

c) The WTO reviews the trade policies of member countries to ensure they comply with WTO agreements.

d) The WTO provides technical assistance and training to help developing countries participate more effectively in the global trading system.

e) The WTO cooperates with other international organizations, such as the World Intellectual Property Organization (WIPO) and the International Monetary Fund (IMF), to promote global economic cooperation and development.

iii. **International Trademark Association (INTA):** International Trademark Association (INTA) was founded in 1878 and is a global association of brand owners and professionals. Its mission is to promote the value of brands and trademarks, and to advance the expertise and professional development of its members. INTA provides educational programs and resources, conducts research and advocacy, and facilitates networking and collaboration among its members. Its structure includes the Board of Directors, various committees and task forces, and staff. INTA has over 7,000 members from more than 190 countries.

iv. **International Association for the Protection of Intellectual Property (AIPPI):** International Association for the Protection of Intellectual Property (AIPPI) was established in 1897 and is an international non-profit organization dedicated to the development and improvement of intellectual property laws. Its mission is to promote and facilitate the international protection of IP, to foster cooperation and communication among its members, and to provide input to international IP policy-making bodies. AIPPI has over 9,000 members from more than 125 countries. Its structure includes the Executive Committee, various standing committees and study groups, and national and regional groups.

v. **International Federation of Intellectual Property Attorneys (FICPI):** International Federation of Intellectual Property Attorneys (FICPI) was established in 1906 and is a global organization of intellectual property attorneys in private practice. Its mission is to promote and advance the practice of IP law worldwide, to provide professional development and support to its members, and to advocate for improvements in IP laws and practices. FICPI provides educational programs and resources, conducts research and analysis, and participates in IP policy-making and advocacy. Its structure includes the Executive Committee, various commissions and working groups, and national and regional sections. FICPI has members from more than 80 countries.

vi. **International Anti-Counterfeiting Coalition (IACC):** International Anti-Counterfeiting Coalition (IACC) was established in 1979 and is a non-profit organization dedicated to combating counterfeiting and piracy. Its mission is to promote and facilitate collaboration among its members, to provide education and training on anti-counterfeiting best practices, and to advocate for stronger laws and enforcement actions against counterfeiting. The IACC includes brand owners, law firms, government agencies, and other stakeholders. Its structure includes the Board of Directors, various committees and working groups, and staff. The IACC has over 250 members from more than 40 countries.

vii. **European Patent Office (EPO):** European Patent Office (EPO) was established in 1977 and has its headquarters in Munich, Germany. Its mission is to grant European patents for inventions, conduct searches and examinations, and provide information and training on IP issues. The EPO is an intergovernmental organization with 38 member states. Its structure includes the Administrative Council, the Examination and Opposition Divisions, and various boards and committees.

viii. **African Regional Intellectual Property Organization (ARIPO):** African Regional Intellectual Property Organization (ARIPO) was established in 1976 under the Lusaka Agreement. It is an intergovernmental organization which provides IP services and protection to its member states. Its mission is to promote creativity and innovation for economic growth and development in Africa. ARIPO aims to provide a regional IP system that facilitates economic growth, technology transfer and encourages research and development. Its membership includes 22 countries from the African region. ARIPO is governed by an Administrative Council and a Director General who is the Chief Executive Officer of the organization.

ix. **Eurasian Patent Organization (EAPO):** Eurasian Patent Organization (EAPO) was established in 1994 by an agreement signed between nine Eurasian countries. Its mission is to promote the protection of inventions and innovation through the provision of high-quality patent services. EAPO's functions include the grant and registration of Eurasian patents, examination of patent applications, and providing patent information services. EAPO currently has eight member states which include Armenia, Azerbaijan, Belarus, Kazakhstan, Kyrgyzstan, Russia, Tajikistan, and Turkmenistan. It is governed by the Eurasian Patent Convention and its supreme governing body is the EAPO Administrative Council.

x. **Intellectual Property Owners Association (IPO):** Intellectual Property Owners Association (IPO) was founded in 1972 as a trade association for companies and individuals interested in intellectual property issues. Its mission is to advocate for strong and effective intellectual property rights protection in the United States and around the world. IPO's functions include providing education and networking opportunities for its members, lobbying for favorable IP policies, and promoting the importance of IP to innovation and economic growth. IPO's membership includes more than 200 companies and organizations from various industries, including technology, pharmaceuticals, and entertainment. IPO is governed by a Board of Directors and several committees and task forces that oversee the organization's activities.

These organizations work towards promoting the protection and enforcement of intellectual property rights globally, and also provide a platform for member countries to collaborate on intellectual property-related issues.

xi. **Trade-Related Aspects of Intellectual Property Rights (TRIPs):** TRIPs stands for Trade-Related Aspects of Intellectual Property Rights, which is an international agreement administered by the World Trade Organization

(WTO). The agreement sets out minimum standards for the protection and enforcement of intellectual property rights (IPR) in WTO member countries. The TRIPs Agreement covers various types of intellectual property, including patents, copyrights, trademarks, geographical indications, and trade secrets.

The TRIPs Agreement includes provisions on the following :

a) *National Treatment:* The agreement requires that WTO member countries provide equal treatment to foreign and domestic IPR holders.

b) *Most-Favored-Nation Treatment:* This provision requires that WTO member countries extend the same level of protection to all WTO member countries.

c) *Minimum Standards of Protection:* The TRIPs Agreement sets out minimum standards of protection for different types of intellectual property.

d) *Enforcement:* The agreement requires WTO member countries to have effective legal remedies for the enforcement of IPR.

e) *Dispute Settlement:* The TRIPs Agreement provides for dispute settlement procedures in case of disputes between WTO member countries over the implementation of the agreement.

xii. **General Agreement on Tariffs and Trade (GATT):** The General Agreement on Tariffs and Trade (GATT) is a predecessor of the World Trade Organization (WTO). It was a multilateral agreement aimed at reducing trade barriers between member countries. The GATT also included provisions on the protection of intellectual property, although these provisions were less detailed than those in the TRIPs Agreement. The GATT provisions on intellectual property were eventually superseded by the TRIPs Agreement when it was introduced in 1995.

Treaties for IPR Protection

i. **Paris Convention for the Protection of Industrial Property (1883):** The Paris Convention for the Protection of Industrial Property was signed in 1883 and is one of the oldest international agreements on intellectual property. Its aim is to protect industrial property, including patents, trademarks, and industrial designs. The Convention provides for the right of priority in the filing of applications for patents and trademarks, meaning that an applicant who files an application in one country has a certain amount of time to file in other countries without losing the priority date. The Paris Convention has 177 member states and is administered by the World Intellectual Property Organization (WIPO).

ii. **Berne Convention for the Protection of Literary and Artistic Works (1886):** The Berne Convention for the Protection of Literary and Artistic Works was signed in 1886 and is the oldest international agreement on copyright. Its aim is to protect literary and artistic works, including books, music, and visual arts. The Convention provides for the automatic protection of works without the need for registration, and sets out minimum standards of protection for authors. The Berne Convention has 177 member states and is also administered by the WIPO.

iii. **Patent Cooperation Treaty (1970):** The Patent Cooperation Treaty (PCT) was signed in 1970 and provides a simplified and centralized system for filing patent applications in multiple countries. The PCT enables an applicant to file a single international application that can be used to obtain protection in over 150 countries. The PCT is administered by the WIPO and has 157 member states.

iv. **Madrid Agreement Concerning the International Registration of Marks (1891):** The Madrid Agreement Concerning the International Registration of Marks was signed in 1891 and provides a system for the international registration of trademarks. The Madrid Agreement allows an applicant to file a single international application that can be used to obtain protection in multiple countries. The Agreement has 107 member states and is administered by the International Bureau of WIPO.

v. **Madrid Protocol (1989):** The Madrid Protocol was signed in 1989 and is an updated version of the Madrid Agreement. It provides a more flexible and efficient system for the international registration of trademarks. The Protocol has 107 member states and is also administered by the International Bureau of WIPO.

vi. **Budapest Treaty on the International Recognition of the Deposit of Microorganisms for the Purposes of Patent Procedure (1977):** The Budapest Treaty on the International Recognition of the Deposit of Microorganisms for the Purposes of Patent Procedure was signed in 1977 and provides for the recognition of the deposit of microorganisms for the purpose of obtaining a patent. The Treaty allows an applicant to deposit a microorganism with a recognized depository, which can be used to support a patent application in multiple countries. The Budapest Treaty has 80 member states and is administered by WIPO.

vii. **Hague Agreement Concerning the International Deposit of Industrial Designs (1925):** The Hague Agreement Concerning the International Deposit of Industrial Designs was signed in 1925 and provides for the international registration of industrial designs. The Agreement allows an applicant to file a single international application that can be used to obtain protection in multiple countries. The Hague Agreement has 74 member states and is administered by WIPO.

viii. **WIPO Copyright Treaty (1996):** The WIPO Copyright Treaty was signed in 1996 and is an international treaty that provides protection of copyright in the digital age. The Treaty sets out minimum standards of protection for authors and provides for the use of technological measures to protect works. The WIPO Copyright Treaty has 98 member states.

ix. **WIPO Performances and Phonograms Treaty (1996):** The WIPO Performances and Phonograms Treaty was also signed in 1996 and provides for the protection of the rights of performers and producers of phonograms. The Treaty sets out minimum standards of protection for performers and producers and provides for the use of technological measures to protect works. The WIPO Performances and Phonograms Treaty has 99 member states.

x. **Nice Agreement Concerning the International Classification of Goods and Services for the Purposes of the Registration of Marks (1957):** The Nice Agreement is an international treaty that establishes a standardized system for the classification of goods and services for trademark registration purposes. It was first adopted in 1957 and has since been revised several times. The main objective of the agreement is to simplify and streamline the registration process for trademarks by providing a clear and consistent classification system that is recognized worldwide. The agreement is administered by the World Intellectual Property Organization (WIPO) and is open for accession to all countries that are members of the Paris Convention for the Protection of Industrial Property.

xi. **Geneva Convention for the Protection of Producers of Phonograms Against Unauthorized Duplication of Their Phonograms (1971):** The Geneva Convention for the Protection of Producers of Phonograms Against Unauthorized Duplication of Their Phonograms, also known as the Phonograms Convention, is an international treaty that provides protection to producers of sound recordings against unauthorized duplication and distribution. It was adopted in 1971 under the auspices of the World Intellectual Property Organization (WIPO) and has been ratified by numerous countries around the world. The mission of the treaty is to ensure that producers of sound recordings are able to control the use and distribution of their works, and to receive fair compensation for their use. The treaty also establishes minimum standards for the protection of sound recordings, including the right to authorize their reproduction, distribution, and rental. Membership is open to all countries that are members of the WIPO or the United Nations.

3

Patents Understanding Essential Concepts for Innovators

To fully harness the potential of human creativity in industrial applications, it's essential to safeguard and organize it effectively. One of the best ways to achieve this is by obtaining a patent. The term "patent" is derived from the Latin word "patere," meaning "to lay open" or make available for public inspection. A patent is an exclusive right granted by the government to an inventor/creator for a limited time in exchange for disclosing their invention for public use. With this right, the patent holder can prevent others from making, using, selling, or exploiting the invention for a certain period, thereby providing commercial benefits to the inventor.

Patents are among the oldest forms of intellectual property rights and are territorial in nature, offering protection only within the country where they are granted. However, through conventions and the Patent Cooperation Treaty (PCT), applicants can seek protection in multiple countries. Obtaining a patent grants the inventor a monopoly right to control the product's production, sale, and marketing. Although the rights granted through a patent are limited to a specific country, extending protection to all contracting countries is possible through a PCT application.

> *A patent is a government-granted exclusive right for a limited time to inventors/creators, in exchange for disclosing their new, useful, and non-obvious invention for public use. This allows the patent holder to prevent others from exploiting the invention for a limited period, granting commercial advantage.*

Evolution of Patent Laws: A Historical Overview

The evolution of patent laws is a fascinating and complex subject that can be traced back to ancient civilizations. The earliest known patent was granted in 500 B.C in Gaurmands city to a confectioner for promoting culinary art on a delicious dish. This practice of granting exclusive rights to inventors expanded to other Greek cities and later to other crafts and commodities.

In 1421, Filippo Brunelleschi was granted the world's first exclusive right over an invention for inventing a method of transporting goods up and down the river Arno in Florence. The modern concept of patents emerged in the 17th century with the introduction of the Statute of Monopolies in England, which provided inventors with exclusive rights to their inventions for a fixed period. From there, patent systems continued to evolve and expand, with many countries implementing their own laws and regulations to protect inventors' intellectual property. Early laws in American colonies aimed to encourage foreign manufacturers to establish new industries by providing them with protected domestic markets. Today, patent laws play a critical role in the global economy, providing inventors with legal protection and a framework to bring their ideas to market and promote innovation.

Evolution of Indian Patenting System

- In 1856, India's first patent legislation, Act VI 1856 on Protection of Inventions, was passed to promote new and useful manufactures and encourage inventors to disclose their inventions. It was repealed by Act IX in 1857 and replaced by a new "exclusive privileges" Act XV 1859, modeled on the UK's 1852 Act, which aimed to promote inventions in India.
- In 1872, Act XIII of 1872 renamed the 1859 Act as "The Patterns and Designs Protection Act," providing protection for designs. It was further amended as Act XVI of 1883, which added a provision for the novelty of the invention, and remained in effect for 30 years.
- The Indian Patents and Designs Act, 1911, replaced all previous Acts in 1911 and was amended in 1920. After Independence, the 1911 Act was amended in 1950 (Act XXXII of 1950) to address the working of inventions and compulsory licenses/revocation under the chairmanship of Justice (Dr.) Bakshi Tek Chand.
- In 1957, the Indian government appointed the Justice N. Rajagopala Ayyanga Committee to revise the Patent Law, which resulted in the Patents Act, 1970, passed on 20th April 1972, with the publication of the Patent Rules, 1972. The Act remained in force for about 24 years without any changes until December 1994, when an ordinance was issued in 1999, which applied retrospectively from January 1, 1995. The amended Act extended patents to drugs, pharmaceuticals, and agrochemicals.
- The second amendment to the Patents Act, 1970, was passed in 2002 (Act 38 of 2002).
- The third amendment to the Patents Act, 1970, came into force on January 1, 2005.

- Currently, India's patent system is governed by the Patents Act 1970, which has been amended several times in 1994, 1999, 2002, and 2005.
- In 2005, the Indian government made significant amendments to the Patents Act to bring it in line with the Trade-Related Aspects of Intellectual Property Rights (TRIPS) agreement. These amendments allowed for product patents in areas such as chemicals, pharmaceuticals, food, and agro-chemicals. In addition, the Patents Rules were also amended several times, starting with the Patent Rules of 2003, followed by amendments in 2005, 2006, 2012, 2013, 2014, 2016, 2017, and 2019. India became a signatory to the Patent Cooperation Treaty (PCT) in 1998, which resulted in a significant increase in patent filings in India, including National Phase applications under PCT.

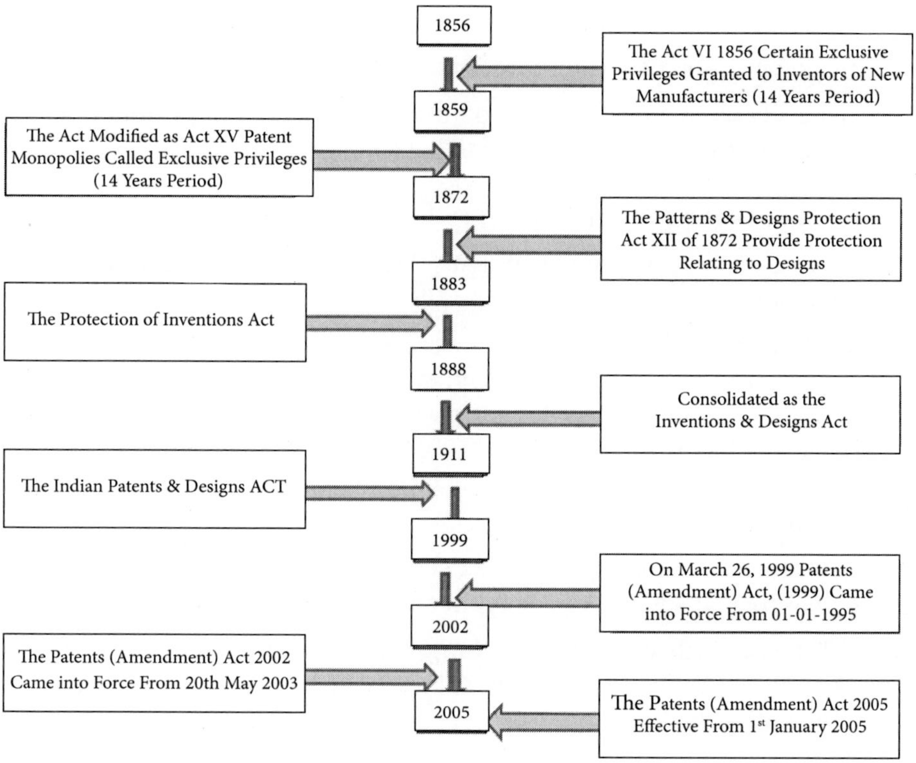

Historical Development of Indian Patent System

What is an Invention

An invention, according to Section 2(1)(j), refers to a new product or process that involves an inventive step and has potential for industrial applications.

Patents are granted for new and useful inventions in various fields such as medicine, industry, chemistry, machinery, and others. For an invention to be eligible for a patent, it must be novel and useful. It's important to note that an invention is distinct from a discovery.

Patentable Subject Matter

An invention can be patented if it involves a new process or product with an inventive step that can be utilized in an industrial application. However, certain categories of inventions are non-patentable as specified in sections 3 and 4 of the Act. The following items are examples of what can be patented:

a) Any article, apparatus, machinery, or its component.

b) Any substance, whether living or non-living, product, or pharmaceutical product.

c) Any composition of matter, including pharmaceutical products.

d) Any process, manner, or art of manufacturing, except for essential biological processes.

Requirements for Patentability

For an invention to be granted legal rights through a patent, it must fulfill three essential requirements:

a) **Novelty:** The product or process must be entirely new and different from prior art. The novelty of an invention should be established at a global level, meaning that it should not have been published previously in any printed or digital media such as magazines, technical bulletins, books, newspapers, or TV.

b) **Inventiveness (Non-Obviousness):** The invention should represent an advancement or modification of existing technology, and it should not be obvious to a person of ordinary skill in that particular field.

c) **Usefulness (Industrial application):** The invention must be capable of industrial application, be useful, fulfill an objective, and not be against public policy. It should provide some benefit or advantage to society.

What are Non-patentable Under the Indian Patent System

Under the Indian patent system, there are certain exceptions to what can be protected by patents. While a range of new products and processes can be patented, the following items are non-patentable in India:

a) Inventions that go against the laws of nature

b) Inventions that are contrary to public ethics

c) Inventions that cause serious harm to human, animal, plant life, or the environment

d) Natural substances such as human genes and new stars

e) The mere discovery of a new property or use for a known substance, or the use of a known process, machine, or apparatus

f) Substances obtained by mere admixture resulting only in the aggregation of the properties of the components or a process for producing such substances

g) The mere arrangement or duplication of known devices functioning independently of one another

h) Methods of horticulture or agriculture

i) Methods of doing medical treatments for humans or animals, or the way in which a disease is diagnosed

j) Plants and animals in whole or any part thereof, including seeds, varieties, and species, and essentially biological processes for production or propagation of plants and animals

k) Mathematical methods and scientific theories, such as Newton's Law of Motion or the Archimedes principle

l) Literary, dramatic, musical, or artistic works or any other aesthetic creations

m) Schemes, rules, and methods used for different things, such as doing a classical dance or running a business

n) Presentations of information

o) Topographies of integrated circuits

p) Inventions that are based on traditional knowledge or an aggregation of known properties of traditional components

q) Inventions relating to atomic energy falling within sub-section (1) of Section 20 of the Atomic Energy Act, 1962

r) Computer programs, although a computer program embedded in hardware may be patentable

s) Combination medicines developed by pharmaceutical companies are not considered to be patentable inventions under Indian law.

Time to File Application

It is recommended to file a patent application as soon as possible without delay. The application can be filed with a provisional application and incomplete specification to establish the priority of the invention. Delay in filing may pose risks such as

i) Another inventor filing a patent application on the same invention and

ii) Inadvertent publication of the invention by the inventor or others independently, which can impact the patentability of the invention.

Important Point Before Patenting an Invention

Before patenting an invention, it's important to keep the following points in mind:

a) Stay curious, but do not disclose your invention to others before filing a patent application.

b) Keep a bound notebook to record the date you came up with the idea, which can be used as evidence in case of a legal dispute.

c) Develop your idea and create a prototype to demonstrate its feasibility.

d) Conduct market research to understand the existing products in the market, identify the need for your invention, and explain what sets your product apart from competitors.

Filing a Patent Application

The process of filing a patent application involves several steps leading up to the grant of a patent. The application should be in the prescribed format and filed with the Controller of Patents. Only one application is allowed per invention, and only one patent can be granted for that invention.

Who Can Apply for a Patent? The application for a patent can be made by the following individuals or entities:

a) **The true and first inventor of the invention,** excluding the first importer of the invention into India or anyone who has received the invention from outside of India. The application must include the name, address, and nationality of the inventor(s).

b) **The assignee of the true and first inventor**, who can be a registered company, research organization, educational institute, or government entity. The assignee must submit "Proof of Right" during the application process. The inventor can assign someone to file the application on their behalf by endorsing Form-1 within 6 months of the application date.

c) **The legal representative** of a deceased person who was entitled to make the patent application prior to their death.

Methods of Filing a Patent Application

There are two ways to file a patent application, either by physical filing at the Patent Office or by electronic filing through the Comprehensive E-Filing portal (https://ipindia.gov.in/e-gateways.htm#comprehensive-e-filing) on the IPR web portal, which was launched on July 20, 2007. The patent application can be filed in either Hindi or English language.

Where to File a Patent Application

The Indian Patent Office has four locations in Kolkata, Delhi, Chennai, and Mumbai, where all patent-related procedures are carried out. The patent application must be filed in the jurisdiction of the respective patent office. The territorial jurisdiction is determined based on the following criteria:

a. The place of residence, domicile, or business of the applicant (first mentioned applicant in the case of joint applicants)
b. The place where the invention actually originated
c. The address for service in India given by the applicant when he has no place of business or domicile in India.

Territorial Jurisdiction of Head office\the branch office to file a patent

Office	Territorial Jurisdiction	Head of the Office, address
Patent Office Branch, Chennai	The States of Andhra Pradesh, Karnataka, Kerala, Tamil Nadu and the Union Territories of Pondicherry and Lakshadweep	Deputy Controller of Patents & Designs and Head of Office Patent Office Intellectual Property Building G.S.T. Road, Guindy, Chennai-600032 Phone: 044-22505242 Fax: 044-22502066 Email: chennai-patent@nic.in
Patent Office Branch Mumbai	The States of Maharashtra, Gujarat, Madhya Pradesh, Goa and Chhattisgarh and the Union Territories of Daman and Diu & Dadra and Nagar Haveli.	Deputy Controller of Patents & Designs, BoudhikSampada Bhawan, Antop Hill, S. M. Road, Mumbai - 400 037. Phone: 022- 24153651, 24148165 Fax: 022-24130387 Email: mumbai-patent@nic.in
Patent Office Branch, New Delhi	The States of Haryana, Himachal Pradesh, Jammu and Kashmir, Punjab, Rajasthan, Uttar Pradesh, Uttaranchal, Delhi and the Union Territory of Chandigarh.	Joint Controller of Patents & Designs Head of Office - Patents Intellectual Property Office Building, Plot No. 32, Sector 14, Dwarka, New Delhi-110078 Phone: +91-11-25300207 Email: delhi-patent@nic.in

Patent Office, HO Kolkata	The rest of India	Deputy Controller of Patents & Designs Intellectual Property Office Building, CP-2 Sector V, Salt Lake City, Kolkata-700091 Phone: 033-23679101, 033-23671987 Fax: 033-23671988 Email: kolkata-patent@nic.in

Types of Patent Applications

There are different types of patent applications that can be filed depending on the nature of the invention and the purpose of the patent. Here are the main types:

1. **Provisional Application:** In India, the "First to File" system is followed, which is also known as the "First-Come-First-Served-Basis." The Provisional Application is filed with incomplete specifications when the invention is still under experimentation and needs more time for finalization. The inventor gets an additional 12 months to finalize the invention with complete specifications. Filing a Provisional Application helps prevent any other related inventions from occurring. Therefore, the applicant who applies first will be granted the patent after examination. It should be filed within 12 months, giving the inventor more time to complete the remaining work. However, to obtain a patent, an ordinary application must be filed with complete specifications.

2. **Ordinary or Non-Provisional Application:** An inventor files this application if they do not have any priority to claim their invention, and it is supported by a complete specification. The work has to be completed in all aspects, and the product/process must be ready to launch in the market. Two types of Ordinary applications are filed with the Controller Office: i. Direct Filing: A complete specification is initially filed with the Indian Patent Office. ii. Subsequent Filing: The application is accompanied by a claim and complete specifications. The application must be filed subsequently to the filing of the corresponding provisional specification. A complete specification includes the following:

    a) Title
    b) Preamble to the invention
    c) Technical field of the invention
    d) Background of the invention
    e) Objects of the invention
    f) Statement of the invention
    g) Brief description of the drawings

h) Detailed description of the invention
i) Claims
j) Abstract

3. **Convention Application:** A convention application allows the applicant to claim priority in all the countries that are signatories to the convention or a group of countries, a union of countries, or inter-governmental organizations. India joined the Paris Convention in 1998, the WTO Agreement in 1995, the Patent Co-operation Treaty in 1998, and the Budapest Treaty in 2001.
4. **Patent Corporation Treaty (PCT) Application:** The PCT is a multilateral treaty that came into force in 1978. India became a member of this treaty on December 7, 1998. Under the PCT, an international system is established to enable the filing of applications in a single patent office in one language. International search reports are generated for invention patentability. PCT also provides international publication for patentability. PCT offers a platform to determine if the claimed invention meets international criteria for patentability and is validated in more than 150 countries worldwide. The priority date in a PCT application is considered in all designated countries.

a) **PCT International Application:** A PCT application is an international application. An applicant from India can file this application at the Indian Patent Office (IPO), which acts as the Receiving Office (RO). The International Bureau of WIPO can receive this application after availing of a foreign filing permit from the IPO or after six weeks and 12 months of filing an application in India.

b) **PCT National Phase Application:** When an international application is made according to PCT, designating India, an applicant can file the national phase application in India within 31 months from the international filing date or the priority date (whichever is earlier). It is necessary for an applicant to file a national phase application in each country where protection is sought. For the national phase application, the title, description, abstract, and claims as filed in the International Application under PCT are considered as the Complete Specification. In addition, the regulations applicable for filing and processing an ordinary patent application also apply here.

5. **Patent of Addition:** This application must be filed if the applicant discovers that they have come across an invention that is a slight modification of the invention that has already been applied for or patented by the applicant. It can only be filed if the invention doesn't involve a substantial inventive step.

6. **Divisional Application:** An applicant may choose to divide an application and file two or more applications if a particular application claims more than one invention.

Patent Specification

Patent Specification is a vital component of a patent application that contains detailed scientific information about the invention, including drawings, and claims to the patent rights. It is a techno-legal document that plays a crucial role in the examination process of the patent application. The specification can be either provisional or complete, and it must be submitted in Form 2 along with the Application Form 1. It is essential to ensure that the specification is written clearly and comprehensively, providing sufficient information to enable a person skilled in the art to understand and replicate the invention. The specification is a critical aspect of the patent application and can determine the scope of the protection granted by the patent.

Filing of Patent Application

The process of filing a patent application involves submitting a single application for one invention using Form-1, accompanied by either a Provisional or Complete Specification in Form-2 and the requisite fee. The standard fee applies for applications with up to 30 pages and 10 claims. Additional fees are charged for applications with more than 30 pages or more than 10 claims. If filing the application in hard copy format, an additional fee of 10% is payable. It is important to ensure that all the necessary documents and fees are submitted accurately and completely in order to avoid delays or rejection of the application.

Documents Required for Filing Patent Application

To file a patent application, one must fill out the prescribed application form (Form 1) with a Provisional or Complete Specification and pay the required fee at the patent office. The application form must be supported by the necessary documents.

i. **Application Form1 Legal Information for Grant of Patent:** The form 1 carried the following information of the patentee.
   - Name and address of the applicant
   - Complete contact details
   - Inventor
   - Assignee
   - A proof of right dully authenticated or Assignments by inventor
   - Details of foreign filing

ii. **Application Form 2:** Application Form 2 contain the technical information (Specification) of the inventions.
   - Background of the invention: Describes the need for the invention
   - Summary of the invention: Describes how the invention works, someone should be able to duplicate your work, the preferred embodiment examples
   - Claims: Most important part, independent claims, dependent claims, Abstract

iii. A statement and undertaking from the inventor in foreign applications under Section-8 (1) in Form-3.

iv. Form-4 is a request to the Controller, for compete specification after one month of application which was not filled in Form 3

v. Form-5 Declaration as to inventorship

vi. The Form 28 accompanied with appropriate evidence in the case of "small entity or startup", If an applicant is MSME/Startup,

vii. Form-26 is a power of authority (if assigned to patent agent) if filed through a Patent Agent Priority document is required in Convention Application (under Paris Convention) and PCT National Phase Application

viii. Form 18 Request for examination-

ix. Form9 Request for Early publication

x. Application should bear the Signature of the applicant or authorized person

xi. The Specification shall be signed by the agent/applicant.

xii. A permission from the National Biodiversity Authority in case of invention related to biological material obtained from India. The specification shall be with source and geographical origin of the biological material.

xiii. Fee as cash/by cheque/by demand draft should be payable to the "Controller of Patents"

Processing of Application

Once the patent application form, along with the necessary documents, is submitted to the patent office, it undergoes several processes before it is finally processed. Firstly, the application is digitized, verified, screened, classified, and uploaded to the internal server of the office. It is then screened for various factors such as International Patent Classification, technical field of invention, relevance to defence or atomic energy, and correctness and completeness of abstract. An examiner is then allocated to the respective field for further examination of the application.

Upon receiving the application, a unique number is assigned to each application in the format of "YYYY J T NNNNNN," where YYYY denotes the year of filing, J denotes jurisdiction, T denotes the type of application, and NNNNNN specifies the continuous running serial number for all patent offices in India. For applications that require expedited examination under Rule 24(C), the numbering format used is "EYYYYJNNNNNN."

Scrutiny of Application

The scrutiny process of a patent application involves assessing important parameters such as jurisdiction, proof of right or power of attorney, signed documents, abstract and drawings, and claims. It should be noted that no patent can be granted for an invention related to atomic energy falling under sub-section (1) of Section 20 of the Atomic Energy Act, 1962.

Request for Patent Examination

The applicant must file a request for patent examination in either Form 18 or Form 18A along with the fee within forty-eight months from the date of priority or filing of the application, whichever is earlier. The examination report is sent to the applicant, and a notice is given to any person interested in the examination.

Patent Application

↓

Publication

↓

Pre Grant Opposition

↓

Examination

↓

Application Grant

↓

Publication of Grant

↓

Post Grant Opposition

Step-by-Step Guide to Obtaining a Patent

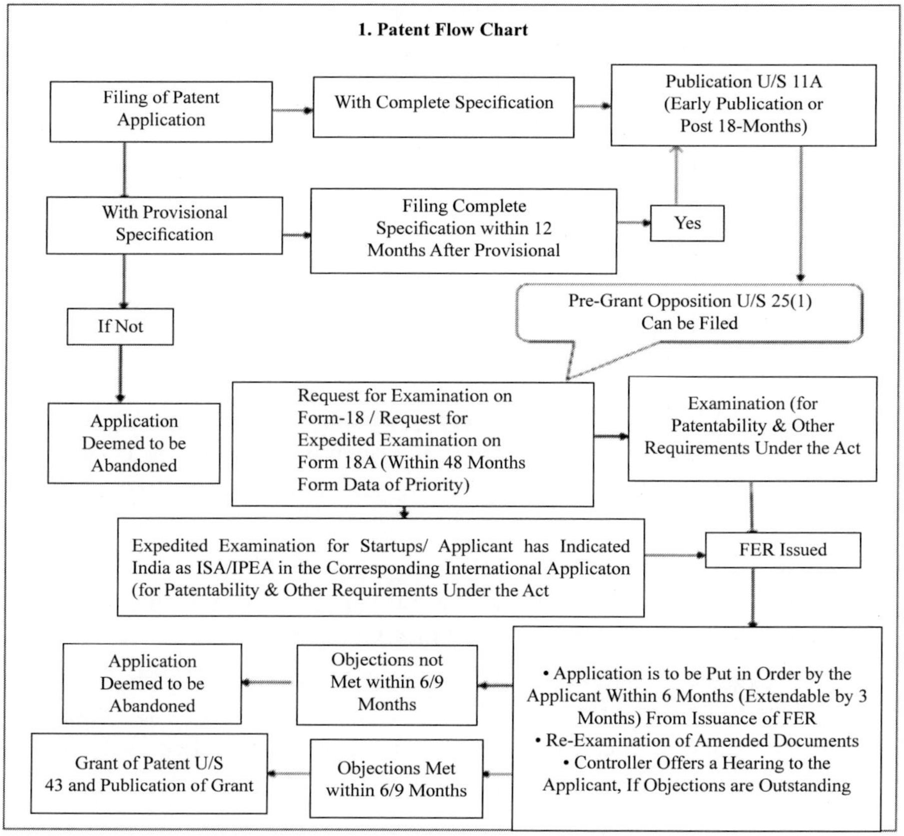

Graphical representation for granting of patent

Duration of Patents

The duration of a patent, as per Section 53, is 20 years from the date of filing the patent application. Once this period ends, the technology becomes part of the public domain, unless the patentee has applied for renewal by submitting the required fee.

Publication of Patent Applications

The patent application is published in the official e-journal at the end of the 18-month period from the date of filing or priority date, whichever is earlier. In some situations, an application may be published upon request made in Form-9 along with the prescribed fee of Rs.2,500/- for natural persons or Rs.10,000 for legal entities other than natural persons.

The publication includes important details such as the application number, date of filing, title of invention, publication date, international patent classification, name and address of the applicant, name of the inventor(s), priority details like priority document number, date, country, etc., reference to Patent of Addition/ Divisional Application along with filing date of the parent application, abstract, number of claims, and drawings (if any).

Once published, patent information is made available to the public on the website, including the abstract, drawings, and any other relevant documents. At this point, the depository institute prepares the biological material for public use.

Withdraw Patent Application

Patent application can be withdrawn in the following ways:

a) A request for the withdrawal of the application can be filed within 15 months of the date of filing or date of priority, whichever is earlier and accordingly the application will not be published and be treated as — withdrawn.

b) An applicant can withdraw the application before the issuance of the First Examination Report. However, on withdrawal of the application, applicant can claim a refund of up to 90% of examination fees.

c) An applicant can withdraw the application any time after its filing but before the grant of patent by making a request. There is no fee for withdrawing the application. And also there is no refund of examination fee.

Patent Opposition and Revocation

Opposition (Right to oppose a granted patent)

Under Section 25 of the Patent Act, any interested person can give written notice of opposition to the Controller against the grant of a patent on valid grounds. This right to oppose a granted patent is available before the grant of the patent, i.e., where an application for a patent has been published, but a patent has not been granted yet.

Person Interested in Opposition

- Person engaged in or promoting research in that field
- Possession of patents related to the same matter
- Manufacturing interest
- Trading interest
- Interest must be real, definite and substantial, and must not arise from what the opponent proposes to do. Thus mere possession of an article covered by the patent is not enough.

- Financial interest
- Specification contains statements that are injurious to the opponent

The process of opposition can be initiated by filling out Form 7 within one year from the date of publication of the grant of patent. The opponent is required to send a written statement to the Controller stating their interest and relevant facts. The Controller must inform the patentee of the receipt of the opposition without delay. The patentee must provide a statement along with evidence within two months of receiving the opponent's written statement. If the patentee fails to reply with evidence within the specified timeframe, the Controller will hear the opponent's case and dispose of the representation in the prescribed manner and time.

To examine the opposition by both the opponent and the patentee, the Controller constitutes an Opposition Board with three members, one of whom acts as Chairman. The Board is required to submit a report within three months from the date on which all relevant documents were forwarded to them. After the hearing, if the granted patent is found to be invalid, the Controller shall issue an order of revocation, and the fact of revocation shall be entered in the register of patents.

Revocation of Patent

Revocation of a patent can be initiated by any interested person or the Central Government by filing a petition before the Intellectual Property Appellate Board. The grounds for revocation are specified under Section 64 of the Patents Act. In addition, a patent may also be revoked by the High Court on a counter-claim in a patent infringement suit.If a granted patent is related to atomic energy, for which no patent can be granted under sub-section (1) of section 20 of the Atomic Energy Act, 1962, the Central Government may direct the Controller to revoke the patent. Furthermore, the Central Government has the power to revoke a granted patent if it is not in the public interest.

Infringement/ patent violation/stolen ideas

Patent infringement occurs when someone uses, makes, sells, or imports a patented invention without permission from the patent holder. This can result in legal action being taken against the infringing party. There are different types of patent infringement, including direct infringement, indirect infringement, and willful infringement.

a) **Direct Infringement:** Direct infringement occurs when someone makes, uses, sells, or imports a product that infringes upon a patent without obtaining a license from the patent holder. In other words, they use the patented invention without permission. The patent holder has the right to file a lawsuit against the infringing party to stop them from using their invention and seek compensation for any damages caused.

b) **Indirect Infringement:** Indirect infringement is when someone contributes to or induces patent infringement. There are two types of indirect infringement: contributory infringement and inducement to infringe a patent. Contributory infringement occurs when someone provides a component or part that is used to make the infringing product, while inducement to infringe a patent occurs when someone actively encourages or assists another party to infringe a patent.

c) **Willful Infringement:** Willful infringement occurs when someone knowingly and intentionally infringes upon a patent. This means that they are aware of the existence of the patent and choose to ignore it. Willful infringement may result in increased damages being awarded to the patent holder in a lawsuit.

Compulsory Licensing

Compulsory licensing is a mechanism that allows governments to grant licenses to third parties to manufacture, use, or sell a patented invention without the consent of the patent owner. This is done when the government believes that the patented product or process is necessary for public health, safety, or welfare, but the patent owner is unable or unwilling to provide the product in sufficient quantities, at reasonable prices, or within an acceptable timeframe.

The Legal Basis for Compulsory Licensing

The TRIPS (Trade-Related Aspects of Intellectual Property Rights) Agreement, which is part of the World Trade Organization (WTO) agreements, allows countries to issue compulsory licenses under certain circumstances, including public health emergencies and situations of national emergency or extreme urgency. The Indian Patents Act, 1970, provides for compulsory licensing under certain conditions, including non-working or insufficient working of a patent, unreasonable pricing, and public interest.

The Process of Obtaining Compulsory Licensing

In India, any person may apply to the Controller of Patents for a compulsory license. The application must state the reasons for seeking the license, including the inability or refusal of the patent owner to meet the demand for the patented product or process. The Controller will examine the application and determine whether the criteria for compulsory licensing have been met. If so, the Controller may issue a compulsory license, subject to conditions such as the payment of reasonable royalties to the patent owner.

Implications of Compulsory Licensing

Compulsory licensing can be a controversial issue, as it may be seen as infringing on the rights of patent holders. However, it is an important tool for ensuring that essential products are available to those who need them most, particularly in developing countries where access to essential medicines, vaccines, and other health products can be a matter of life and death. By allowing generic manufacturers to produce patented products, compulsory licensing can help reduce the cost of these products, making them more accessible to the poor and marginalized.

Grounds for Compulsory Licensing

Under the Patents Act, 1970, there are several grounds on which a compulsory license can be granted. These include:

1. **National Emergency:** If the government considers it necessary for the public health or national security, a compulsory license can be granted.
2. **Public Non-commercial Use:** A compulsory license can be granted if the patent owner has not made the invention available to the public on reasonable terms, or if the invention is not being worked in the country.
3. **Anti-competitive Practices**: If the patent owner is engaging in anti-competitive practices, such as price gouging or abuse of market power, a compulsory license can be granted.
4. **Export of Patented Products:** If the patented product is being exported to another country where it is sold at a lower price than in the home country, a compulsory license can be granted.
5. **Failure to Meet the Demand:** If the patent owner is unable to meet the demand for the product, a compulsory license can be granted.

Procedure for Granting Compulsory Licensing

The process for granting a compulsory license involves several steps:

1. **Application:** The applicant must file an application with the Controller of Patents, stating the grounds on which they are seeking a compulsory license.
2. **Notification to the Patent Owner:** The Controller of Patents must notify the patent owner of the application.
3. **Opportunity to be Heard:** The patent owner has the right to be heard and to provide evidence to support their case.

4. **Decision:** The Controller of Patents must consider all the evidence presented and make a decision on whether to grant the compulsory license.
5. **Terms and Conditions:** If a compulsory license is granted, the Controller of Patents will determine the terms and conditions under which the license will be granted.
6. **Compensation:** The patent owner is entitled to compensation for the use of their patent. The amount of compensation will be determined by the Controller of Patents.

Compulsory licensing is a powerful tool that can be used to ensure that essential products are available to those who need them most. However, it is important to balance the needs of the public with the rights of patent holders. By following the proper procedures and considering all the evidence, a fair and equitable decision can be reached.

Some Examples of Compulsory Licensing Issued in India

- Nexavar (Sorafenib tosylate) - a cancer drug manufactured by Bayer was issued compulsory licensing by the Indian Patent Office in 2012. Natco Pharma was given permission to manufacture and sell a cheaper generic version of the drug.
- Imatinib Mesylate - a cancer drug manufactured by Novartis was refused a patent by the Indian Patent Office in 2006. In 2013, the Indian Supreme Court upheld the decision and allowed Indian pharmaceutical companies to manufacture and sell generic versions of the drug.
- Tenofovir Alafenamide Fumarate (TAF) - a drug used for the treatment of HIV/AIDS was issued compulsory licensing by the Indian Patent Office in 2019. Hetero Drugs was given permission to manufacture and sell the drug at a lower price.
- Dasatinib - a cancer drug manufactured by Bristol-Myers Squibb was issued compulsory licensing by the Indian Patent Office in 2013. BDR Pharmaceuticals was given permission to manufacture and sell a generic version of the drug.
- Sunitinib Malate - a cancer drug manufactured by Pfizer was issued compulsory licensing by the Indian Patent Office in 2012. Cipla was given permission to manufacture and sell a generic version of the drug.
- Darunavir - a drug used for the treatment of HIV/AIDS was issued compulsory licensing by the Indian Patent Office in 2014. Hyderabad-based Hetero Drugs was given permission to manufacture and sell the drug.

- Ibrutinib - a cancer drug manufactured by Janssen was issued compulsory licensing by the Indian Patent Office in 2017. Natco Pharma was given permission to manufacture and sell a generic version of the drug.
- Velpatasvir - a drug used for the treatment of Hepatitis C was issued compulsory licensing by the Indian Patent Office in 2017. Natco Pharma was given permission to manufacture and sell the drug.
- Saxagliptin - a drug used for the treatment of Type 2 Diabetes was issued compulsory licensing by the Indian Patent Office in 2019. AstraZeneca was directed to grant a license to Torrent Pharmaceuticals.
- Enzalutamide - a cancer drug manufactured by Astellas Pharma was issued compulsory licensing by the Indian Patent Office in 2018. BDR Pharmaceuticals was given permission to manufacture and sell a generic version of the drug.
- Lenalidomide - a cancer drug manufactured by Celgene was issued compulsory licensing by the Indian Patent Office in 2019. Natco Pharma was given permission to manufacture and sell a generic version of the drug.
- Rivaroxaban - an anticoagulant drug manufactured by Bayer was issued compulsory licensing by the Indian Patent Office in 2019. Natco Pharma was given permission to manufacture and sell a generic version of the drug.
- Trastuzumab - a cancer drug manufactured by Roche was issued compulsory licensing by the Indian Patent Office in 2013. Biocon was given permission to manufacture and sell a biosimilar version of the drug.
- Sitagliptin - a drug used for the treatment of Type 2 Diabetes was issued compulsory licensing by the Indian Patent Office in 2013. Merck was directed to grant a license to Glenmark Pharmaceuticals.
- Sorafenib Tosylate and Erlotinib Hydrochloride - cancer drugs manufactured by Natco Pharma were issued compulsory licensing by the Indian Patent Office in 2010. The company was given permission to manufacture and sell generic versions of the drugs.

Why Compulsory Licensing Common in Medicines and Drugs in India?
Compulsory licensing is more common in medicine and drugs in India due to the importance of access to affordable healthcare for a large population that may not be able to afford the high costs of patented medicines. India is also known for its generic drug industry, which produces low-cost versions of patented drugs and exports them to developing countries.In the case of life-saving drugs, such as those used to treat HIV/AIDS, cancer, or hepatitis, the

high cost of patented drugs can make them inaccessible to many patients who need them. Compulsory licensing can provide an avenue for the government to ensure that these essential drugs are available at affordable prices.Moreover, the Indian Patent Act, 1970, allows for compulsory licensing in certain situations such as national emergencies, public health crises, or when the patent holder is found to be abusing their patent rights. This makes it easier for the government to grant compulsory licenses for drugs that are deemed to be of public importance, without violating the TRIPS agreement.

List of Patent search Resources/Patent Databases in Different Countries

Country/Organization	URL
Australia	http://www.ipaustralia.gov.au/patents/search_index.htm (Patents)
Benelux	http://register.boip.int/bmbonline/intro/select.do?language=en (Trademarks)
Canada	http://brevets-patents.ic.gc.ca/opic-cipo/cpd/eng/introduction.html (Patents) http://strategis.ic.gc.ca/cipo/trademarks/search/tmSearch.do (Marks)
Chile	http://www.dpi.cl/dpi_web/Frm_Login_default2.htm (Patents & Trademarks)
China	http://www.sipo.gov.cn/sipo_English/zljs/ (Patents)
Czech Republic	http://isdvapl.upv.cz/pls/portal30/pt.ptfrm (Patents) http://isdvapl.upv.cz/pls/portal30/oz.ozfrm (Trademarks)
European Patent Office	http://www.espacenet.com (Patents)
Finland	http://www.prh.fi/en/patentit/tietokannat/patinfo.html (Patents)
France	http://www.inpi.fr/front/show_rub.php?rub_id=8 (Patents)
Germany	http://www.dpma.de/suche/suche.html (Patents)
Hungary	http://www.hpo.hu/English/adatbazis/ (Patents)
India	http://www.pfc.org.in/db/db.htm (Patents)
Ireland	http://www.patentsoffice.ie/eregister/Query/PTQuery.asp (Patents)
Japan	http://www.jpo.go.jp/quick_e/index_search.htm (Patents & Trademarks)
Latin America	http://invenes.oepm.es (Patents)
LUSOPAT (Portal of Portuguese language IP collections)	http://www.portal-lusofonia.org
Mexico	http://siga.impi.gob.mx (patents information search) http://vidoc.impi.gob.mx (electronic files search)
New Zealand	http://www.iponz.govt.nz/search/cad/dbssiten.main (Patents & Trademarks)